baby OSTRICHES

KIM THOMPSON

CREATIVE EDUCATION • CREATIVE PAPERBACKS

CONT

CONTENTS

I AM A CHICK.

I am a baby ostrich.

My wings are
not for flying.
Ostriches are
flightless.
wing
two-toed foot
with claws

My dad dug a nest in the ground.
My mom laid eggs inside.

After 40 days, I <u>hatched</u>. In just a few days, I was ready to walk away from the nest.

I live in Africa with my [herd]. We walk across the grassland. We look for leaves, seeds, insects, and lizards to eat.

The sun is hot. I stay in my mom's shadow.

By the time I am two, I am grown.

If danger comes, I fight back. My kick is strong enough to kill.

I can run as fast as a car!

SPEAK AND LISTEN

CA-CA-CAAWW!

Can you speak like a chick?
Baby ostriches chirp and trill.

Listen to these sounds:

https://www.youtube.com/
watch?v=cZ00FltTmwc

Now it is your turn!

CHICK WORDS

beak: the horny, pointed jaw of a bird

flightless: not able to fly

hatched: broke out of an egg to be born

herd: a group of animals that travel together

READING CORNER

McGrath, Barbara Barbieri. *Five Hiding Ostriches.* Watertown, Mass.: Charlesbridge, 2022.

Poliquin, Rachel. *Ostriches (Superpower Field Guide).* New York: Clarion Books, 2020.

Riggs, Kate. *Ostriches (Amazing Animals).* Mankato, Minn.: Creative Paperbacks, 2023.

INDEX

**PUBLISHED BY CREATIVE EDUCATION
AND CREATIVE PAPERBACKS**

P.O. Box 227, Mankato, Minnesota 56002
Creative Education and Creative Paperbacks
are imprints of The Creative Company
www.thecreativecompany.us

**LIBRARY OF CONGRESS CATALOGING-
IN-PUBLICATION DATA**

Names: Thompson, Kim, 1970- author
Title: Baby ostriches / Kim Thompson.
Description: Mankato, Minnesota : Creative Education
   and Creative Paperbacks, [2026] | Series: Starting out
   | Includes bibliographical references and index. |
   Audience term: juvenile | Audience: Ages 4-7
Creative Education and Creative Paperbacks |
   Audience: Grades K-1 Creative Education and
   Creative Paperbacks | Summary: "Introduce beginning
   readers to the world of baby ostriches with this life
   science starter. Includes photos, a labeled animal
   diagram, "Make a Noise" section, glossary, and further
   resources"-- Provided by publisher.
Identifiers: LCCN 2024043243 (print) | LCCN 2024043244
   (ebook) | ISBN 9798889897514 library binding | ISBN
   9781682778371 paperback | ISBN 9798889897644 ebook
Subjects: LCSH: Ostriches--Infancy--Juvenile literature
Classification: LCC QL696.S9 T46 2026 (print) | LCC
   QL696.S9 (ebook) | DDC 598/.51--dc23/eng/20250107
LC record available at https://lccn.loc.gov/2024043243
LC ebook record available at https://lccn.loc.
   gov/2024043244

DESIGN AND PRODUCTION

Design by Rhea Magaro
Production by Beeline Media and Design, Inc.
Art direction by Tom Morgan

PHOTOGRAPHS by Alamy Stock Photo/Ann and Steve
Toon, 2-3, Scott Hurd, 9; Dreamstime/Gueret Pascale, 6-7,
Iuliia Ariukh, cover, Natalia Golovina, 11; Getty Images/
Mike Powles, 10-11; Shutterstock/ABCDstock, 11, Bohbeh,
13, Eric Isselee, 5, GeorgiaaaB, 7, IZAN IS, 12, Jacky
Photographer, 4, nadi555, 14, Toomko, 8

Printed in India